PROPOLIS

This book explains the medical significance of propolis – a resinous substance gathered by bees from the leaf buds and bark of trees and used to maintain and disinfect their hives. Includes suggested treatment and dosage for such ailments as throat infections and coughs, ulcers, wounds, gum disorders and halitosis.

PROPOLIS
The Natural Antibiotic

by

RAY HILL

NATURE'S WAY

THORSONS PUBLISHERS LIMITED
Wellingborough, Northamptonshire

First published 1977
Seventh Impression 1986

ISBN 0 7225 0353 9

Printed and bound in Great Britain

CONTENTS

INTRODUCTION

It has been said that propolis is the most effective natural antibiotic yet discovered by man, and the strange thing is that its discovery probably took place over 2,000 years ago. So much ancient wisdom has been lost to modern man, but fortunately this remarkable substance, put into daily use by the bees, has in recent years been 'rediscovered' and its effectiveness is constantly being proved by scientific research.

Propolis is just one of the products resulting from the ingenious organization of the beehive and is of enormous benefit to man. It is a resinous substance, gathered by the bees from the leaf buds or bark of trees, particularly poplars, and used as a 'cement' in the maintenance of the hive. Bees, themselves amazingly complex creatures in their way of life, seal up any cracks and holes in the hive and fix the comb to the roof with this, their own brand of building material, and they also take advantage of its antibacterial qualities by using it to encase any unhygienic foreign bodies that get into the hive but which they cannot remove – cocooned in propolis, any decomposing matter that threatens to pollute the bees' living quarters is made safe.

This instinct to plaster everything with propolis was a great source of irritation to beekeepers in the past, but nowadays, as research into the therapeutic effects of this substance widens, it has become one of the most valuable products of the hive, for the implications to be drawn from the properties it contains are indeed exciting.

Before the rediscovery of propolis, there was one disadvantage to the Nature Cure methods of diet, water cure, fasting and exercise, and this was the fact that there was nothing to replace the allopath's antibiotic as a speedy way of alleviating and curing bacterial ailments. The naturopath's treatment, although one which helped the whole body to a better state of health rather than merely supplying drugs for a particular problem, required time before the real benefits could be seen.

AN ALTERNATIVE TO ANTIBIOTICS

Propolis now offers an alternative to

antibiotics in that it is immediate in its action, yet has none of the side-effects that drugs can produce, and it is in this role as nature's non-toxic antibiotic that fascinating results are emerging.

Although it has proved effective for a variety of ailments, it is in the treatment of throat infections that propolis has shown its most remarkable results. I myself have given raw propolis chips to people with sore throats and coughs, and almost without exception the symptoms have disappeared within hours. The taste of propolis in this crude state may not be to everyone's liking, but it is now marketed in the form of chewy lozenges with a fruity flavour, and these work just as well.

There is also evidence that it relieves halitosis and gum disorders, acne and other skin complaints, ulcers, wounds, and various other disorders, details of which are given in the last chapter of this book. And as these things are discovered, so propolis is being sold in forms that make it more effective for a particular condition; for example, there is propolis toothpaste for gum disorders and halitosis, capsules which dissolve in the stomach for internal problems, ointment for skin troubles, and tincture for wounds and so on, or as a gargle to soothe sore throats.

So, as can be seen, propolis is making an exciting comeback into the field of natural medicine, and as its rediscovery is still in its infancy, whole areas of its possible powers remain as yet untapped. Many people will not even have heard of it, yet old herbals prescribe it for numerous ailments, and its antibacterial properties have been acknowledged since the beginning of recorded history. Why, although the other products of the beehive have continued to rank as some of the most effective health-giving substances available to man, propolis has had this period of obscurity is mysterious, but the worldwide interest now being shown in it by researchers is fast bringing it back into the limelight, and I hope that this book will help to spread the word even further.

WHERE PROPOLIS FITS IN

To look at the hive as a whole and study the life-style of the bee is interesting, for almost every part of its system produces something that is of use to man.

Honey is of course the most well-known product, and one which has very real healing effects. The worker bees spend much of their lives gathering nectar and pollen, which is stored in the cells of the comb, and it takes something like two million bee flights from hive to plant to make one pound of honey.

It contains vitamins of the B complex and a significant amount of vitamin C, and is also an excellent source of minerals such as calcium, copper, iron, magnesium, manganese, potassium and sodium, as well as of protein. One of the most easily assimilated foods, it is recommended for a variety of ailments and as an important food supplement. But another very important attribute of honey is its antiseptic property, proved time and time again by bacteriologists and medical scientists; indeed, in some hospitals it is used as a dressing for wounds after surgical operations. It also makes a fast-healing salve for burns, and there is nothing new-fangled about any of these discoveries for in the fourth century BC, Hippocrates, the Father of Medicine, prescribed honey for sores and ulcers.

Honeycomb, which includes the wax manufactured from the bees' wax glands to make up the cells, has long been considered to be one of the finest remedies for hay-fever, but it is now thought that honey cappings, the thin wax cappings that are sliced from the comb before the honey is extracted, provide an even better one. People have found that if they chew honey cappings for a month before the hay-fever season begins, they will not suffer attacks at all, or at worst only extremely mild ones. Severe sufferers, however, would need to chew honeycomb or honey cappings three times a day, starting three or four months before the hay-fever season and continuing the treatment into it if necessary.

It is thought that the minute amounts of pollen in the comb and cappings serve to make this such a successful remedy. The body thus has a chance to develop enough antibodies to ward off the attacks which normally occur during heavy pollen counts.

Research has proved that pollen itself is one of the richest sources of vitamins, minerals, fats, enzymes, and hormones, as well as having a very high protein and amino acid content. The bee itself feeds on pollen and until recently man had to depend upon the industry of the bees to collect pollen for human use – nowadays, however, it is harvested by machines direct from the plant. It is a strangely underestimated food supplement, for a dose of high-potency pollen every day can improve the state of health to such an extent that many athletes maintain that it has considerably enhanced their performance. Older people, particularly, find it beneficial, and it is a specific remedy for depression, fatigue, jaded nerves, and many medical problems such as anaemia, disorders of the colon, and as a restorative for those recovering from serious illness or shock.

When the honeycomb is constructed in the hive, some of the cells are made to slightly larger dimensions, and these are known as the 'royal cells'. The eggs which are laid in these cells by the queen bee are designed to produce potential queens, for the grubs will be fed on royal jelly, with which the royal cells are provided. Royal jelly is a viscid substance secreted by the glands near the mouth of the honey bee, and contains many vitamins of the B complex, including pantothenic acid which has now been proved to be efficacious in the treatment of arthritis. This is the food which gives the queen bee the potential for an amazingly prolific output during her egg-laying period. From a human point of view, it is recommended as an energy-giving food and also has rejuvenating qualities, as well as having been found excellent in the treatment of some heart conditions.

Even the wax produced by the bees is useful to man, though not as a food or medicine, since as a furniture polish beeswax has no equal. And even the thing which makes most people steer clear of bees – their sting – has in several cases proved to have a marked therapeutic effect in the treatment of arthritis and rheumatism, and is used by some practitioners as a regular form of treatment for these conditions.

And so the phrase 'a hive of activity' was well coined, for as we shall see these treasures of the hive have been produced by ceaseless industry. The intricate and beautifully symmetrical construction of the comb gives us honeycomb and honey cappings; the endless flights out to plants, collection and storage in the cells give us honey and pollen, the production of food for the royalty of the hive provides us with royal jelly, and the collection from leaf buds of a substance with which to maintain the hive and keep it free from bacteria is man's source of propolis.

It is with the last of these, not only from a therapeutic point of view, but in its context as a vital link in the activity of the hive, that this book is concerned.

CHAPTER ONE
THE BEE'S ANTIBIOTIC

Just as the construction industry needs cement for building, so the bee needs something of a similar nature to fix the honeycombs to the hives and to stop up any cracks and crevices, and as is to be expected, the ingenuity of the bee in providing itself with a suitable substance is as resourceful as man's own. The raw material the bee uses is a resinous substance exuded by the buds of trees such as the horse-chestnut, and found in cracks in the bark of trees such as spruce, larch and other conifers, but especially in the leaf buds of poplars. This substance is propolis, and it is far more than a building material alone.

It has been found that the bee produces several substances with antibacterial properties, but propolis is one of its most important agents against infection in the hive, and whether by its scent, or by instinct, they will soon discover the best source of propolis in their neighbourhood. Bees live at very close quarters – a hive may contain as many as 40,000 or 50,000 of them – so that unchecked infection could swiftly spread to cause a large-scale disaster.

The word propolis comes from two Greek words meaning 'defences before a town', and the appropriateness of its name is illustrated by the barrier of propolis which bees sometimes build behind the entrance to the hive, so that all the inmates must pass through it on their way in and out. It is indeed a vital element of their activities for they will use it as a defence against any potential disintegration of the hive on one hand and for the removal of the possibility of pollution from foreign bodies on the other. It points out the remarkable organization of nature that the bees have managed to discover such a useful substance and integrate it so ingeniously into their way of life.

WHAT PROPOLIS CONSISTS OF

Propolis is a sticky substance which protects leaf buds and prevents them from drying out, and it varies in colour according to the plant source – it may be dark brown, or a lighter shade, and can even be of a reddish or violet hue.

Although   its   biologically   active

components may vary according to its source, a test of propolis collected in fifteen different districts of the USSR showed uniform constituents, the approximate breakdown of which was 50-55 per cent resin and balsam, up to 30 per cent wax, about 8 or 10 per cent fragrant essential oils and about 5 per cent solid matter. It is said to be rich in fats, amino acids, organic acids, composite ethers of univalent alcohols, and trace elements such as iron, copper, manganese, zinc and others, tannic acid, phytoncides, and antibiotics. Apart from this, it has a high vitamin content, especially those of the B group, but also E, C, H, P, and protovitamin A, because pollen accounts for 5-10 per cent of its composition.

Further analyses show a formidable list of strange-sounding ingredients including cinnamic acid; cinnamyl alcohol; vanillin; chrysin; galangin; acacetin; kaempferid; rhamnocitrin; pinostrobin; caffeic acid; tetochrysin; isalpinin; pinocembrin; and ferulic acid.

ANTIBIOTIC PROPERTIES

The antibiotic properties of propolis are believed to come from the flavanoids it contains – particularly galangin, the name of which comes from the active substance in galingale, an aromatic plant root from the East which is related to ginger and has always been used medicinally as well as for culinary purposes. Plants containing flavonoids had been used as natural remedies for centuries before these active substances were scientifically identified, but in the late twenties, a Hungarian, Szent-Gyorgi, isolated vitamin C from oranges, and the discovery of flavonoids stemmed directly from this.

Vitamin C (ascorbic acid) is known even to those who have never studied nutrition, but less known is Szent-Gyorgi's discovery in 1936 of a substance in lemon peel which was effective in cases of bleeding which had not responded to the administration of vitamin C alone. This substance, called citrin, is also made up of flavonoids, and its efficacy appeared to come from a strengthening effect on the capillary walls of the blood vessels. The chief flavonoid in citrin is hesperidin; another named rutin was found in tobacco and buckwheat plants. The flavonoids were called vitamin P, and it was found that a combination of vitamin C and vitamin P had in many cases a dramatic effect on a variety of disorders including blood vessel ailments, some types of haemorrhages, and even virus infections and rheumatism. The action of flavonoids is not precisely understood, but it is believed to lie partly in the prevention of rapid oxidation of vitamin C, resulting in the strengthening of the body's own defences against disease and infection. The antibiotic properties of the flavonoids in propolis have a similar effect.

PREPARATION FOR HUMAN USE

Most of the propolis for human use comes from poplars which are found in the woods and forests of central Europe, although, as has been mentioned, other trees also produce the substance. Its preparation for human consumption is subject to strict controls, and each batch is analysed to ensure it is free from any contamination and is genuine propolis manufactured by the bees from plant resin.

In its raw state it looks like mineral chippings, because it is literally chipped away from the hive and breaks off in little pieces. It is sometimes sold in this form, or is ground and marketed as a powder, loose or in capsules. In order to make it yet easier to take and more practical for the consumer, it is also available as an ointment or cream for both oily and dry skins, in tincture form for internal and external use, and is incorporated into lozenges and toothpaste for a number of mouth infections.

EXPERIMENTS AND TESTS

A great deal of work has gone into finding the best methods of preparing propolis for human use and in establishing how it will keep, as well as whether the preparations remain active in storage. The French entomologist, Dr P. Lavie, described how the extract was made by heat treatment in either alcohol or water: 50 grammes of the propolis was boiled for an hour in 1 litre of each solvent in a flask with a condenser. The extract was then filtered, evaporated in a water bath and redissolved. He found that in most tests the alcoholic extract proved slightly more active than the aqueous one. The extracts redissolved in water were stable for several months when kept in a refrigerator and protected from light. Russian workers, however, considered that long storage (three or four years) did not reduce the content of the chemical components of propolis extracts, and did not result in reduction of antibacterial acti ty.

In other Russ an experiments, extracts were prepared by mixing one part by weight of propolis with two parts alcohol. The mixture was left to stand for three to four days, during which time it was transferred frequently from one vessel to another. After filtering through muslin, the residue was weighed to ascertain the concentration of the extract. The required concentration was then adjusted by adding alcohol, the solution obtained being a brown colour. To produce an alcoholic watery solution, the necessary quantity of distilled water was then added to the extract.

In Germany, Kohler extracted propolis with a dilute aqueous solution containing water soluble organic compounds if desired. Filtration and acidification of the extract gave a preparation which, after decanting, washing and vacuum drying, was a pale yellow to brown amorphous material with a strong odour. Small amounts of wax were removed by treatment with carbon tetracholoride and the material was purified chromatographically.

It is through these various experiments and tests that it has become possible to make propolis available in different forms to match specific requirements and to make taking it as beneficial to the taker as possible.

CHAPTER TWO

A SALVE KNOWN TO THE ANCIENTS

Man's interest in bees and his dependence on them for sweetness in his food goes back as far as the records of human life itself, and many legends in which these insects feature have been handed down to us from the ancients. One of the oldest rock paintings, in the Cuevas de la Arana in Valencia, shows a hole in a cliff with bees flying around and two men climbing primitive ropes to take the honey. Bees were also widely depicted on tombs, coffins and vases from ancient Egypt, while the sign of the bee was associated with the titles of the kings, and was used as the motif on ornaments presented as rewards for valour.

By the time of the Greek and Roman authors, the art of beekeeping had been established. Virgil, for example, was a beekeeper, and wrote extensively on the subject, both in practical and poetic terms, and some of the observations made by early writers on the life of the bee are astonishingly accurate, particularly as they had no scientific means of checking their discoveries.

MYTHS AND LEGENDS

One area in which knowledge was lacking, however, was the generation of bees themselves. It was widely believed that they obtained their young from flowers, particularly the olive, and the best-known legend of all was that bees were born from the carcase of an ox. These insects were probably drone flies, and it is thought that they deceived Samson just as they deceived Virgil and others. When an ox was killed, it was lain on a bed of thyme and sealed up in a room for three weeks. The room was then opened to the air, and some eleven days later there would be nothing left but a skeleton and clusters of bees.

Legend also has it that Jupiter transformed the beautiful Melissa into a bee, and that bees were bred from hornets and the sun.

In Great Britain the custom of 'telling the bees' still persists in country areas – the bees must be 'told' or languish and die. Swarming, too, is regarded as lucky in some parts of the world and unlucky in others;

Virgil, for instance, considered a swarm to be an unlucky omen – he thought it foretold the coming of a foreign army and its general. Bees have always been regarded as weather prophets, with the result that honey was often used in rain magic. It has also been used in both birth and death rituals – the Assyrians and Egyptians buried their dead in wax and honey, an early use of the antibacterial qualities of the hive.

FOLK MEDICINE AND PROPOLIS

But it is not only legend that surrounds the bee, for the medicinal properties of the products of the hive have been known and used for centuries. Books about folk medicine from all over the world show that since the beginning of recorded history resins have been used in wound dressings and to heal inflammations and infections. Without being able to analyse their effects, man found that these natural substances would form a protective coating over his wounds, would draw out foreign bodies and help promote healing. Resin preparations were also taken internally for stomach and urinary disorders.

In his massive *Natural History*, Pliny (first century AD) mentioned the uses of resins in general, but went into considerable detail when dealing with propolis. He stated that there were three separate layers in the cementing materials used by the bees.

First they construct combs and mould wax, that is, construct their homes and cells, then produce offspring, and afterwards honey, wax from flowers, bee-glue from the droppings of the gum-producing trees – the sap, glue and resin of the willow, elm and reed. They first smear the whole interior of the hive itself with these as with a kind of stucco, and then with other bitterer juices as a protection against the greed of other small creatures, as they know that they are going to make something that may possibly be coveted; with the same materials they also build wider gateways round the structure.

The first foundations are termed by experts *commosis*, the second *pissoceros*, the third *propolis*, between the outer cover and the wax, substances of great use for medicaments. Commosis is the first crust, of a bitter flavour. Pissoceros comes above it, as in laying on tar, as being more fluid than wax. Propolis is obtained from the milder gum of vines and poplars, and is made of a denser substance by the addition of flowers, and though not as yet wax it serves to strengthen the combs; with it all approaches of cold or damage are blocked, and besides it has itself a heavy scent, being in fact used by most people as a substitute for galbanum.

It seems probable that all three layers were in fact propolis, but obtained from different sources. In another volume Pliny referred to the current medicinal uses of propolis, saying that it 'extracts stings and all substances embedded in the flesh, reduces swellings, softens indurations, soothes pains of the sinews and heals sores when it seems hopeless for them to mend.

'THE BEST EXTRACTIVE'

Celsus, in the first century AD, wrote:

> The following mature abcessions and promote suppuration: nard, myrrh, costmary, balsam, galbanum, propolis, storax, frankincense, both the root and the bark, bitumen, pitch, sulphur, resin, suet, fat, oil. ... The best extractive, however, is that called by the Greeks *rhypodes*, from its resemblance to dirt. It contains myrrh, crocus, iris, propolis, bedelium, pomegranate heads, alum, mistletoe juice, turpentine resin or he-goat's suet.

Dioscorides wrote in similar vein of:

> ... the yellow bee-glue that is of a sweet scent and resembling styrax, is to be chosen, and which is soft in ye excessive dryness of it, and easy to spread after the fashion of mastic. It is extremely warm and attractive, and drawing out of thorns and splinters. And being suffumigated it doth help old coughs and

being applied it doth take away the lichens. It is found about ye mouths of hives, in nature like unto wax.

THE BRITISH HERBALS

In John Gerard's famous herbal *The Historie of Plants* (1597) reference is made to 'the rosin or clammie substance of the blacke Poplar buds ... ' and the fact that it was used by apothecaries to make ointments. And Nicholas Culpeper's *Complete Herbal*, under the heading 'The Poplar Tree' states 'The clammy buds here of, before they spread into leaves, are gathered to make Unguentum and Populneum ... '. Culpeper goes on to say 'The ointment called Populneum which is made of this Poplar, is singularly good for all heat and inflammations in any part of the body and tempers the heat of the wounds'.

In later herbals, poplar resin was acknowledged as having considerable medicinal value, but mostly in the form taken directly from the tree and not from the beehives. In Green's *Universal Herbal* (1824) two poplar species are listed. Under *Populus Nigra* (Black Poplar Tree) we read:

The young leaves are an excellent ingredient for poultices for hard and painful swellings. The buds of both this and the White Poplar smell very pleasantly in the spring, and being pressed between the fingers, yield a balsamic resinous substance which, extracted by spirits of wine, smells like storax. A drachm of this tincture in broth is administered in internal ulcers and excoriations and is said to have removed obstinate fluxes proceeding from an excoriation of the intestines.

Another interesting use is mentioned under *Populus Balsamifera* (Common Tacamahaca Poplar Tree):

The buds of this tree from autumn to the leafing are covered with an abundance of a glutinous yellow balsam, which often collects into drops, and is pressed from the tree as a medicine. It dissolves in the spirits of wine; and the inhabitants of Siberia prepare a medicated wine from the buds. This wine is diuretic, and, as they think serviceable in the scurvy.

It can be seen, therefore, that resins were used extensively in the treatment of wounds and inflammations, often in conjunction with honey, and propolis, according to the Roman author, Merula, was even then being bought by the doctors for a price higher than that of wax. Centuries later, the wheel may well have turned full circle!

CHAPTER THREE
THE BEE'S STORY

The activities of the beehive are fascinating in their ingenuity, and propolis, as we have seen, plays an important role, both from a constructional and a hygienic point of view.

When a queen with her colony enters a fresh hive, the worker bees at once begin to clean and refit their quarters. The propolis 'foragers' set out to collect their building material and then begin the intricate work of using it to cement any cracks or minute holes. While they are busy on this miniaturized housework, other bees hang from the roof in a solid curtain and manufacture from their wax glands the wax for making the comb. The comb is made up of six-sided cells and is suspended from the top of the hive; these cells are so incredibly regular in shape and size that it was once even suggested that they could be used as a unit of measurement!

In these beautifully constructed cells, the queen lays her eggs at the astonishing rate of over 3,000 a day. Some of these will be laid in the 'royal cells' which are larger than others in the comb and are supplied with royal jelly. From these will emerge potential queen bees. Royal jelly, a glandular secretion produced by the worker bees, is a highly concentrated food and although all the grubs in the hive, it is thought, are fed on it for the first two or three days of their existence, only the potential queens continue on this diet. The workers, produced from fertilized eggs laid in the smaller cells, and the drones, from unfertilized eggs, are then fed by 'nurse bees' with pollen and honey. During this egg-laying marathon, the queen's attendants feed and clean her, and worker bees fly out in search of nectar and pollen for storage in the cells.

THE BEE EMERGES

The larva spends about three weeks in its cell, emerges from a chrysalis and, by eating its way out through the mixture of pollen and wax which caps the cells, feeds itself for the first time. Worker bees soon instinctively join in the 'chores' of the hive, gathering nectar, pollen or propolis and helping with cleaning, feeding and repair work. The drones are lazy and have to be fed; their sole purpose in life is to fly out and mate with the young queens from other hives. After this mating flight they die – any which are left alive at the end of the season are killed by the workers.

The newly emerged queen bees are all potential rivals to the existing queen. She will leave, taking a swarm of workers with her, and fly to a new hive, when a young queen will take over the old hive. If two young queens leave their cells at the same time, they will fight to the death, for there can only be one ruler in the hive. The ruling queen in a hive is looked after by the inmates until she is no longer able to lay a steady supply of eggs. When her strength begins to fail she will be put to death, either by the workers themselves or by a younger, emerging queen.

WORK IN THE HIVE

So the world of the bee is a microcosmic civilization. Worker bees forage for food and propolis, feed the embryos, feed and groom the queen, and even have a system of cooling the air in the hive by the rapid fanning of their wings. They clean the hive, taking out any foreign matter, but anything which is too large for them to remove is sealed over hygienically with propolis. Mice, for instance, will occasionally get into a hive. The bees can sting it to death, but cannot remove the corpse, so they cover it with a bacteria-proof 'skin' of propolis, and it thus will not contaminate their surroundings. During the long winter, when the bees cluster together in the hive for warmth, this cleanliness will keep them free from infection.

But, certainly before the medicinal properties of propolis were rediscovered, the hygienic habits of the bee were regarded as a nuisance by most beekeepers. One story tells of an ingenious arrangement of drawers invented by an enterprising beekeeper to facilitate the gathering of honey. All this inventiveness came to nought because the bees simply glued all the drawers together at every joint with propolis! It was also regarded as a contaminant of the wax, for it spoiled it and made it char and clog if used for candles. The dealer would test the beekeeper's wax and if it contained very much propolis he would have to accept lower payment.

Nowadays, however, propolis is regarded as a valuable asset for, on a weight-for-weight basis, it is the most expensive item to come from the hive.

COLLECTING THE PROPOLIS

Although it is easy to see how the bee uses propolis in the hive, it is only recently that the actual collection of the substance has been investigated in detail. The activities of the bees have been documented more thoroughly than those of any other insect, and gradually the areas of doubt have been explored and mysteries solved – from the generation of the bee which puzzled the ancients, to its mating flight, and even the discovery of which colours bees are able to distinguish – because propolis is taken from the tops of trees, its collection has been the most difficult of the bee's activities to observe.

Some of the most detailed information available comes to us because of the persistence and ingenuity of a German scientist, Waltraud Meyer, of the *Zoologisches Institut der Freien Universität*, Berlin, who carried out his observations by setting up artificial sources of propolis. He took some from the hive and placed it in a dish in a selected place, which gave him excellent opportunities for studying the bees at work. This could not of course exactly duplicate natural conditions, as the propolis from the hive was more solid than the sticky covering of the tree buds, but he could at least examine the method used.

Collection under natural conditions begins at the end of June, with marked activity in

the late summer and autumn, and finishes
when the weather becomes too cold for flying
in October or November.

The bees are particularly active on hot
days as the sunshine makes the raw propolis
more workable – it becomes softer and
breaks more easily. But Meyer found that
bees would occasionally collect in very
unsuitable weather – windy, wet, and quite
cold.

BRINGING IT HOME

The bees seem to have two ways of
transporting propolis to the hive, depending
upon the distance involved. If collected far
from the hive, it is packed into loads in the
bee's *corbiculae* (pollen baskets). But when
Meyer put it on the alighting board outside
the hive it was gripped with the mandibles
and taken inside in small lumps. An
interesting detail in his observations was that
if there was a gap of only 1.5 centimetres
between the source of propolis and the
alighting board, the bee would still pack the
propolis in its corbiculae!

Meyer was able to break down into steps
the extraordinarily deft manner in which the
bee handles and transports its sticky load. If
the propolis is solid it is nibbled off, but if it is
soft, as on a hot day, the bee grips it with the

mandibles, then moves its head backwards, pulling the substance into a long thread which finally breaks off. The two forelegs reach forward to knead and shape the lump, after which they take it from the mandibles. It is then passed, with intricate movements by one of the middle legs, to the corbicula on the same side. While thus engaged, the bee is already groping with its antennae for more propolis.

USING THE PROPOLIS

Once back in the hive, the bee waits near the scene of cementing activity while other bees come and remove the propolis from the load, particle by particle, and deposit it where it is needed. Meyer recorded that it took between one and several hours for a propolis forager to get rid of its load in this way.

The remarkable smoothness of the cementing is not achieved by any 'polishing' action, but simply by the continuous nibbling off of minute rough particles.

Meyer also made observations on the work force of the bees, and found that two groups were engaged in cementing. There were 'cementing' bees who kept strictly to their own work, and 'casual workers' who helped wherever they were needed. By marking individual bees, he was led to believe that although all propolis foragers also did cementing work (usually carried out later in the day), not all cementing bees foraged for propolis.

He also discovered that the foragers could easily be diverted into collecting honey or sugar syrup by placing a container on top of, or near to, the propolis. The bee approached to take the propolis, found the syrup and took that back to the hive instead. Arriving for the next load, it would go straight back to the syrup and ignore the propolis altogether. If the syrup was removed after the bee had made several trips, it would search for it for five or ten minutes and then go back to regular propolis collecting. This behaviour was observed with twenty-six bees, who all reacted in the same way, and led Meyer to wonder if propolis was only collected in the late summer and in the autumn because there was so little nectar available.

CHAPTER FOUR
MEDICAL USES

Although propolis has had a period of relative obscurity as regards medical uses, it was still being applied to slow-healing wounds during the Boer War, and even during the Second World War in Russia. After 1945, little attention was paid to the study of natural medicinal substances in the West, for resources were being concentrated on the development of chemically synthesized drugs, but in the USSR far greater importance is attached to this subject than it is in the West, and extensive research into the properties of propolis have been carried out there. But researchers in different countries, working independently and without intercommunication, will sometimes come to similar conclusions at about the same time, and it was because of this that interest in propolis revived in the West in the 1950s.

RESEARCH AND ITS FINDINGS
The most important European discoveries

about propolis were made almost
accidentally, and the facts are set down in a
doctoral thesis by Dr Lavie.

'The study of antibiotics in bees (*Apis
Mellifica L.*) arose by chance,' he wrote.
'Some freshly killed bees were put into a
liquid culture medium without any aseptic
precautions, yet no bacterial development
resulted.'

It therefore became clear that the bee was
equipped with an extraordinarily powerful
weapon in the shape of a natural antibiotic.
A series of experiments followed in an
attempt to identify this substance, but Dr
Lavie was simply studying the natural
biochemical defence of insects rather than
looking for a new antibiotic. It had already
been established by White in 1900 that
material found in beehives was remarkably
free from bacteria and that honey and bee
larvae were always sterile, but the discovery
of antibiotics by Fleming was yet to come,
and at this stage work progressed no further.

Lavie's experiments showed that in the bee
and its surroundings could be found at least
seven different antibiotics. The first was the
one which had originally attracted his
attention, a substance from the body of the
bee itself, and the others were found in the
glandular secretions of worker bees, in wax,

pollen, honey, royal jelly, and propolis.

ACTIVITY OF PROPOLIS ANTIBIOTIC

It has already been mentioned that when the antibiotic extract from propolis used in experiments was prepared in both alcoholic and aqueous solutions, the alcoholic extract in most cases proved slightly more active than the aqueous one. The activity of the propolis antibiotic against several different types of bacteria was compared with that of the antibiotics taken from other hive products. The propolis preparation showed 'interesting activity on *B. subtilis* (Caron.), *B. Alvei* and *Proteus vulgaris*, was less active against *Salmonella pullorum*, *Salmonella gallinarum*, S. type Dublin, *Escherichia coli*, B., and *Bacillus* larvae'. It was not active against four strains of *Escherichia coli* and *Pseudomonas pyocyanea*. Of the substances tested, propolis was the only one which was also shown to have fungicidal properties.

Lavie became interested in the fact that samples of propolis did not all have a constant antibiotic potency, and considered that this could be explained by the fact that the samples often came from different sources. He differentiated between the propolis obtained directly from the outside of buds or trees, and that manufactured by the bees from the resinous substances present in pollens. As poplar trees are the most important source of propolis, Lavie referred to the substance named chrysine which occurs both in propolis and in the poplar buds themselves, together with the leaves and green parts of the tree. He made extracts from the buds and found that their antibiotic action was almost identical with that obtained from propolis when tested on seven different kinds of bacteria. Extracts obtained from other trees gave very variable results, and none was as potent as that from *populus nigra*. It is not surprising, therefore, that propolis is always thought of in connection with poplars.

PRACTICAL PROOF

Laboratory tests will obviously have less meaning for the layman than practical demonstrations. There are records, however, of practical tests with propolis which are of interest. For example, in the same way that propolis will kill harmful bacteria, it can also stop growth in the form of plant germination. Gonnet found that potato tubers placed in occupied hives failed to germinate as a result of substances inhibitory to plant growth, which were deposited on them by the bees. When the coating was sufficiently thick, tubers removed from the hive showed permanent inhibition, and of the hive products examined, only propolis caused this effect.

Rumanian workers also found that alcoholic extracts of propolis at 1:10 dilution inhibited the germination of hemp seeds. An even more practical instance was given by Kivalkina, namely the fact that pieces of meat embedded in propolis are preserved because all microbes which could cause decay are killed – the meat will therefore keep its colour, odour and consistency for a long period of time.

MEDICAL USES
As far as the direct medical uses of propolis are concerned, reports have been collected from all over the world, and some of the most interesting are summarized here.

Inflammations of the Throat and Mouth
Dr Maximillian Kern, of the Clinic at Ljubljana in Yugoslavia, established very good results in cases of inflammation of the mucous membranes of the throat and mouth by giving such patients propolis bonbons to chew. In cases of acute inflammation, almost all patients were free of fever and felt no pain on swallowing as little as six to ten hours

after starting the treatment. The method used was to dissolve one propolis lozenge in the mouth at two-hourly intervals until the temperature had returned to normal and swallowing was painless. Even patients suffering from chronic inflammation of the mouth and gums found that symptoms were hardly noticeable by the following day.

Halitosis

Dr Kern added that propolis was also given to patients suffering from bad breath, and that symptoms had entirely disappeared after a few days. After two months he checked all the patients again and found no return of the condition, and also that there was no adverse reaction to propolis, as there is with most antibiotics.

Tonsillitis

Also from Ljubljana comes the report of a case of severe and persistent tonsillitis in a four-year-old girl, the daughter of a dentist. The child had been ill for some time, and when seen by the doctor had a temperature of 39.7°C (103-5°F), was drowsy and refused food. The doctor prescribed tincture of propolis on sugar lumps, which could be sucked easily. After two doses, the child slept well, her temperature had dropped to 37.6°C (99.7°F), and her appetite had returned. Subsequent examination by the ear, nose and throat specialist showed the tonsils clean and free from inflammation.

Stomach Ulcers

Ulcers feature very largely in reports on propolis from all over the world. Dr F.K. Feiks, of the Public Hospital at Klosterneuberg in Austria, used it in treating both resident patients and out-patients. Fifteen out-patients with proven ulcers were treated *exclusively* with tincture of propolis. Only one case subsequently required hospital treatment; the other fourteen remained in their own homes and the ulcers healed. By comparison, a further seventeen out-patients were treated with conventional medicaments, and of these, eleven had to be hospitalized later as the ulcers caused serious complaints or could not be healed over a long period.

With in-patients, Dr Feiks used propolis tincture as a supplementary therapy in 108 cases out of a group of 294, the remaining 186 remaining as controls. After two weeks, over 90 per cent of the propolis patients were free of symptoms against 55 per cent of the control patients, and in addition the number of operations necessary during hospitalization was reduced by one-third. Dr Feiks stressed that when patients remained under observation for at least two years, relapse remained equal in both groups, so that while the activity of the ulcer is eased, a chronic ulcer condition will tend to remain. Dr Feiks considered that observation would have to be extended over a longer period in order to establish if a repeated prophylactic treatment at critical times would achieve a cure. He cited one instance which gave grounds for hope – that of a female patient of eighty-one who had had a chronic stomach ulcer for twelve years. She had not been operated on because of a heart condition, and annual X-ray controls had always shown the ulcer to be unchanged. She was given a six-week course of treatment as an out-patient, using propolis tincture alone, and the ulcer was subsequently found, by X-ray examination, to have healed. In this case it did not reappear later, and when the patient died of a stroke at the age of eighty-five, the post-mortem showed only the scar.

In Yugoslavia, the case is recorded of a fifty-year-old mechanic, a heavy smoker, suffering from stomach ulcers. He had severe pain after every meal, and in an attempt to avoid this he began to eat less and less. The result was considerable weight loss, leading to weakness and reduced efficiency in his work. He was recommended to try propolis capsules three times daily, half an hour before meals. From the very first day he was without pain, and soon found that by continuing to take the capsules regularly he was able to eat anything and gradually began to regain weight.

Burns

In the USSR in 1958 Demecky recommended propolis ointment for second-degree burns; its efficacy resulted apparently from the tannin content of the propolis, the cleanliness of the wound surface and the soothing changes of dressing which the ointment made possible. The anaesthetic properties of propolis are also valuable in the case of burns, and, furthermore, healing is effected without disfiguring scars.

Dermatology

A great many conditions, both mild and serious, are grouped under this term, and a high proportion of them may be alleviated by the use of propolis.

Striking results have been reported from Austria in the treatment of acne cases. Dr Edith Lauda tested propolis tincture and ointments in the treatment of fifty-nine patients who had suffered for several years from acne of varying degrees of severity, and which had withstood treatment in dermatological clinics. Previous treatments had included antibiotics taken internally,

together with the external application of
cortisone and other ointments. The
conditions treated extended from simple
comedone acne to *acne pustulosa* and *acne
conglobata*.

Dr Lauda reported that twenty-five cases
of *acne simplex* were completely healed by
home treatment with propolis tincture and
ointment within a week. Thirty-five cases of
acne simplex combined with *acne pustulosa* were
healed by home treatment in three weeks,
with only three weekly treatments at the
clinic.

Among the most notable results was the
dramatic improvement in a woman who had
been treated unsuccessfully for thirty years
for *acne conglobata* on the chin. After only two
treatments at the clinic the infiltrated parts
of the skin were free of inflammation and
only small remainders of acne were visible.
Another woman, aged forty, had *acne
pustulosa* covering her whole face, and had
unsuccessfully tried every available therapy.
Here, too, acne was eliminated within two
weeks by home treatment with tincture and
ointment. Dr Lauda pointed out that the use
of propolis carried no danger to the system.
Internal findings, particularly regarding the
ovaries and the gastro-intestinal tract, were
neutral, and no organic or neuro-vegetative

disturbances of any kind were established.

A report from the USSR states that in cases of neurodermatitis and dry eczema there was diminution or complete cessation of itching after twice-daily application of a lanolin, vaseline and propolis ointment, and this was followed by a complete cure of the condition. The report adds, however, that wet eczemas were aggravated by the propolis ointment. It was curative in strepto-dermatitis, but not in staphylodermatitis.

Slow-healing Wounds

Propolis has been found to have a stimulating effect on the regenerative processes of the skin, as it promotes wound granulation and is believed to have antiphlogistic, or cooling, qualities. When mixed into an ointment with vaseline it has been used successfully on slow-healing war wounds; other suitable bases include lanolin and sunflower oil. The use of propolis had been found particularly suitable following amputations.

In Yugoslavia a case was reported of a fifty-four-year-old miner whose right ear was amputated after the discovery of a malignancy. The operation was successful, but the patient requested that he be provided with an artificial ear for cosmetic reasons. The artificial ear was later torn off in an accident and the resulting wound became infected. As the area had previously been subjected to powerful irradiation, the hospital had to carry out a skin transplant, but after a few days the wound began to suppurate. The patient was allowed to go home, but had to return to the hospital regularly for the wound to be cleaned. After a year it seemed unlikely that new skin would ever grow around it.

When the man later visited an otologist because of an inflammation in the auditory passage, the doctor, on learning the patient's history, cleaned the wound surrounding the ear and treated it with lanolin and propolis ointment. He continued this treatment twice a week, and when healing began he substituted an application of propolis tincture with a covering of gauze to encourage drying. During the treatment there was no sign of infection or suppuration. The skinless area became smaller and smaller and after two months it was healed, completely clean and of a normal colour.

Ear Infections

Many other cases have been reported of the favourable effects of propolis on the auditory passages. A forty-year-old woman, a diabetic, was found to be suffering from eczema and inflammation of both auditory passages. Cortisone treatment was effective, but the symptoms returned after treatment ceased. The otologist changed the treatment, using a propolis ointment every second day. Healing began after a week, the irritation in the ears diminished, the surface of the skin lost its scaly appearance and improved in colour, and the patient's hearing also improved. It is well known that diabetics are especially sensitive to infections, and suppurations will often occur with skin troubles. The rapid healing which resulted was, therefore, all the more noteworthy.

Besides inflammations of this type, propolis has also been used – notably in Russia – to treat some types of hearing defects. A 30-40 per cent alcoholic tincture of propolis was mixed with olive oil or maize oil (1:4) to make a creamy emulsion. Patients with various ear diseases and defects were treated by inserting a gauze plug, soaked in the emulsion, into the aural passage. In adults, the plug was left for between thirty-six and thirty-eight hours and the treatment repeated ten or twelve times. Out of 382 patients treated, 314 showed improved hearing. Fewer patients reported head noises after treatment than previously.

Anaesthetic Effects of Propolis

In Russia in 1955, Prokopivic established the

anaesthetic quality of propolis solution by carrying out trials on rabbit cornea. An 0.25 per cent propolis solution was 3.5 times more effective than a corresponding cocaine preparation and fifty-two times more effective than a novocain preparation in the same concentration. It is considered suitable for anaesthetic use in some nasal operations, especially if the patient has a sensitivity to other anaesthetics. It has also been used in dental practice and for the anaesthesia of the gums and mucous membranes of the mouth for small surgical operations. In 1973 it was reported from Russia that a new anaesthetic preparation had been developed using a mixture of novocain and propolis.

Since 1953 the following preparation has been used in dental practice in the USSR: 2-4 per cent alcoholic solution of propolis (40 grammes dry propolis in 100 millilitres 70 per cent alcohol), left for three days, shaken occasionally, then filtered through dense gauze.

Other Indications

The uses of propolis range from the treatment of corns – long-standing in folk medicine – to present-day reports of its protective effects against radiation. It has been successfully employed after tonsillectomy, when it was used to staunch blood, its glutinous composition having a varnishing effect on the operation wound. It has been used in the form of eye drops to reduce intra-ocular pressure. A Soviet author reported the successful treatment of bone joint tuberculosis with local applications of ointment and with a propolis-butter extract taken orally. Some types of virus influenza will also respond favourably to treatment with propolis.

ALLERGY TO PROPOLIS

Although propolis is a natural and harmless substance which can be of great benefit, as is the case with most substances, a small number of people are allergic to it, and it goes without saying that anyone who develops a rash after using it either orally or externally should discontinue the treatment. Not surprisingly, the few cases recorded in detail concern beekeepers who come into constant contact with the substance during the summer months. In 1967 a paper was published at the Department of Dermatology, Royal Infirmary, Edinburgh, on *Contact Dermatitis in Beekeepers due to Propolis*.

The report estimated that possibly 0.05 per cent of beekeepers in Britain might be affected. A typical case history was that of a sixty-two-year-old man who had been keeping bees for thirty years. During the previous six years he had developed an itchy rash affecting the chin, neck, face, hands, and wrists after working in the apiary. The attacks lasted from one to two weeks and were serious enough to keep him off work. His rash appeared only during the times when he was handling bees, honeycombs or frames. There was a row of poplar trees near his hives, as well as some willow and spruce trees, which the bees would have visited for propolis. At the hospital a patch test was made to the beekeeper's own propolis and it was strongly positive.

A survey was subsequently made by the department, when beekeepers who claimed to have a rash after handling propolis were contacted. Similar case histories were collected and it was notable that in all cases the skin remained clear during the winter months when no work was done on the hives.

Incidentally, beekeepers may find a silicon barrier cream helpful, as it prevents the propolis from sticking to the skin during handling and makes it much easier to wash off.

In previous cases reported from the Continent it had been discovered that some patients who reacted to propolis were also sensitive to poplar resins, other tree resins and balsam of Peru. It was suggested that cinnamic acid derivatives might be the cause of the allergy, but the report from Edinburgh stated that the strong sensitizer in propolis and in poplar was apparently none of the cinnamic derivatives they had tested up to the time of reporting. The author of this report remarked that there were far more poplar trees in the British Isles than was commonly realized and that the sudden sensitization of a beekeeper after thirty or forty years was sometimes due to the recent planting of poplars within the two or three miles flight range of his bees.

CHAPTER FIVE
HOW TO USE PROPOLIS

Propolis, as we have seen, is an age-old remedy that is once again gaining recognition and so, because people may not know how best to use it medically and in what dosage (for, as it has been classed as a food supplement, there will be no instructions on the packaging), I have listed a number of complaints where either my own experience or that of others has indicated that a specific dose has produced a beneficial result.

If a particular ailment is not mentioned, it does not necessarily mean that propolis cannot be used in its treatment. But since the recent researches began, there has not been time to build up evidence for every illness and it is thus not yet possible to substantiate all findings. For instance, although rheumatism does not seem to respond to propolis in the cases tried, it may well be that it does in others; and though it has value in the treatment of some forms of arthritis, it does not follow that it will help in every case.

Work is being carried out in Scandinavia into whether propolis is effective in the treatment of some tumours, but again the results remain to be seen. This goes to show, however, that tests and experiments are going on all the time and we will continue to learn more and more about propolis.

PREPARATIONS

The first thing to ascertain is the various forms in which propolis can be bought. Although until quite recently only *raw propolis chips* were available, the taste of these does not appeal to everyone, and other methods of taking propolis have now been devised. The *propolis lozenge* is a delicious sweet and is really effective for sore throats and coughs. *Propolis capsules*, which are hard gelatine capsules containing finely ground propolis, are very useful for stomach and intestinal infections.

Propolis cream is available for both dry and oily skins, and this should obviously be used as an external ointment. If care is taken in choosing the right one for the type of skin, excellent results can be expected. Propolis cream can also be used as a cosmetic for difficult skin.

The name of *propolis skin tonic* can cause

confusion, for although it is a good tonic for the skin, it is in fact a tincture of propolis, and as such can be used internally, either as a gargle (four or five drops in half a glass of warm water) or a few drops can be put on a sugar lump and sucked for throat infections. This particular skin tonic is sold under the brand name Salvaskin, and as far as I know it is the only true tincture of propolis. Therefore, when tincture of propolis is prescribed for a particular ailment in this chapter, Salvaskin Propolis Skin Tonic is what the reader wishing to obtain it should ask for.

Propolis toothpaste is also available now and is especially beneficial for those suffering from gum disorders. It should be used as a regular tooth-paste whether there is any infection or not, however, because the antiseptic, antibacterial action of propolis will kill off the harmful bacteria which breed in the spaces between the teeth, so preventing gum disorders from developing.

All these products are available in health food stores and can sometimes be found in chemists' shops which specialize in health foods and allied products.

INFECTION OF URINARY TRACT

For infections of the kidney, bladder,

prostate gland and sexual organs, take 3 grammes of propolis for the first three days and 2 grammes for about eight days thereafter.

INFECTION OF DIGESTIVE TRACT
Infections of the digestive tract are known to succumb very quickly to propolis chewed raw or in lozenge form for the upper tract, and in powder (capsule) for the stomach and intestines. This should be taken for about five days, after meals in doses of 2 grammes spread over the day. Chronic infections will obviously take longer to cure.

SPECIFIC DOSES
Propolis can be tried for anything in which harmful bacteria, germs or viruses are involved. If a positive sign is not seen within three weeks in acute cases, then it may be assumed that the treatment is not going to work. In chronic cases, however, it can take far longer, and a course of treatment should last at least eight weeks.

The following alphabetical list of ailments is designed to guide the reader in the correct use of propolis.

Abscess
Apply *tincture of propolis* to the affected part.

Acne

All forms of acne benefit from treatment with
propolis. Apply *tincture of propolis* daily until
the condition subsides. Further applications
of *propolis cream* may be used from time to
time as a preventative where this condition is
of a long-standing nature.

Bad Breath (Halitosis)

Suck a *propolis lozenge* every two or three
hours, or as required.

Bleeding

Because of its glutinous nature, propolis is a
most useful remedy for staunching bleeding,
and for this tincture of propolis should be
used. Alternatively, *raw propolis chips* can be
chewed and the saliva applied to the wound.

Corns

Propolis cream used on corns is an age-old
remedy. Apply night and morning and cover
with a small gauze pad.

Coughs

A *propolis lozenge* should be sucked as often as
required, or *raw propolis* may be chewed from
time to time. A gargle made from *tincture of
propolis* should be taken on rising and
retiring: prepare this by placing four or five
drops of the tincture in half a glass of warm
water. The gargle may be swallowed if
desired.

Cuts

Because of its antibacterial properties, *tincture of propolis* should always be used on cuts to reduce the risk of infection.

Cystitis

Cases of cystitis have been greatly improved by taking one *propolis capsule* three times daily. Indeed, propolis can be tried in all conditions affecting the urinary system.

Eczema (Dry only)

The application of *propolis cream* once daily can have an outstanding effect on this form of eczema. It is also recommended that *propolis capsules* be taken twice daily for a few days at the beginning of the treatment.

Propolis can aggravate Wet Eczema, however, so it should never be used for this condition.

Gum Disorders (Gingivitis, etc.)

Take one *propolis lozenge* three or four times daily or chew a small amount of *raw propolis* from time to time. The teeth should always be cleaned with *propolis toothpaste*.

Psoriasis

Follow the instructions given for Dry Eczema. This is a difficult disease to treat, so if there is no visible sign of improvement after a week there is little value in carrying on with the treatment, and further advice should be sought.

Shingles

Take one *propolis capsule* twice daily between meals, and apply *propolis cream* to the affected part before retiring. This treatment has even proved to be efficacious when the disease has been present for many years, as it has the effect of eradicating the itching which makes this particular ailment so irritating.

Sinusitis

As with all infections of the mucous membranes, propolis often has a rewarding effect here. Chew *propolis lozenges* or *raw propolis* as often as required.

Sore Throat

As I have said earlier in this book, it is with this condition that propolis can have the most amazing results. In many cases two or three *propolis lozenges* can clear all the symptoms of a sore throat. Suck a lozenge every hour for up to four hours, continuing the treatment until there is no pain on swallowing.

Gargling with a solution of four or five drops of *tincture of propolis* in half a glass of warm water is also very soothing for a sore throat.

Teeth

As I have already mentioned, *propolis toothpaste* has a healthy effect on the gums and helps to eliminate harmful bacteria. It will

also keep teeth white and is recommended
for regular use.

Tonsillitis

Suck four or five drops of *tincture of propolis* on
a lump of sugar three times daily, or suck a
propolis lozenge three or four times daily.

Toothache

Because of its anaesthetic effects, propolis
can be applied to the region of the aching
tooth until such time as the dentist can be
visited. Use *tincture of propolis* for this – put a
few drops on a pad of gauze and apply to the
affected part. Alternatively, chew *raw propolis*
until it has the consistency of chewing gum
and then let it adhere to the gum around the
tooth.

Ulcers

External. Apply *tincture of propolis* to a piece of
gauze and place it over the ulcer. The sting
this causes can be reduced by diluting the
tincture slightly with boiled water. *Propolis
cream* applied to the dry edges of the ulcer as
the area decreases will help it to heal.

Internal. Propolis is a valuable remedy for
stomach and intestinal ulcers. The best
treatment is with *propolis capsules* – take one
capsule three times daily, thirty minutes
before meals.

Mouth. Apply *tincture of propolis* to the affected
part two or three times a day. *Propolis lozenges*

may also be used.

Wounds (Fresh and Slow-healing)

Tincture of propolis should be applied to wounds and as they heal and dry out *propolis cream* can be used. Diabetics should take heed of this since there is often a tendency for skin infections to occur, and it is well known that their wounds take longer to heal.

Two important points bear repetition at this stage:

1. The tincture of propolis mentioned in this chapter is Salvaskin, and as far as I know, at the time of writing, this is the only pure tincture on the market in Britain. Obviously, however, other brands may emerge, and there are no doubt different types outside the UK. If the reader is offered another brand, he should check very carefully that it is pure tincture before taking it internally.

2. About 0.05 per cent of the population can react adversely to propolis – they are allergic to it, just as people are allergic to a variety of things. This allergy takes the form of a rash, which disappears as soon as treatment with propolis is stopped. If you have not taken propolis before, it is therefore as well to check whether you come into this category. Before going to

bed, take a very small amount and if the skin remains clear, the treatment can begin. However, if there are signs of a rash, then propolis is not for you and should not be taken in any form.

But propolis can and will be of great benefit to many people, and I hope that this book will help to spread the word about the remarkable powers of this natural antibiotic.

INDEX